Numbers! Numbers!

By
K Stoughton, M. Ed.
Aka Auntie G 123

ISBN: 979-8-9885503-1-0
KS Publications & Ed., LLC

Revised edition

Numbers do important jobs. They help us in lots of ways. Let's learn them and use them to help us.

Symbols save us time.

Symbols can be a fast way of telling someone something. A symbol is a picture, icon or shape that stands for something else. Have you seen these symbols in the world? Do you know what they mean?

The $ is for money – in this case dollars. Numbers, like 2 or 3, are special symbols that stand for how many.

If we didn't have numbers, then …

How do we know how many crackers we have?

How can we tell which player is which on our favorite sports team?

$$$ And very important, how do we know how much money we have or how much apple juice costs?

Zero has an important job.
It means none or not any.

0 cookies

$ 0

This symbol can be very confusing because it is also used as a letter to make words. You have to look closely at what's around the 0 to know what it means.

Zero has more jobs than we have room for in this book.
We will use another book to learn to tell you about them.

1 sock

1 banana

$ 1

The $ means dollars which is money. Isn't that symbol a good shortcut?

2 socks

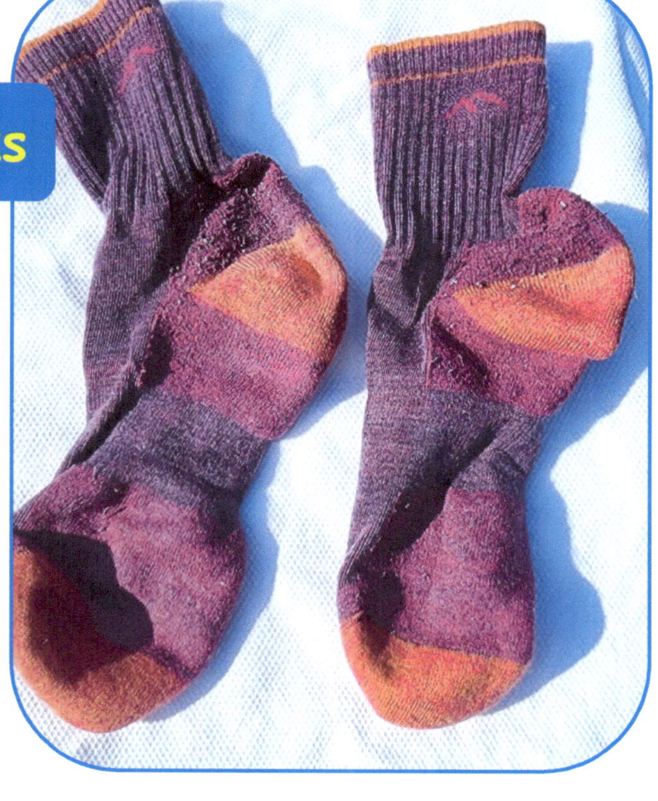

2 chicken tenders
Yum!

$ 2

Can you count 3 lighted dragonflies?

$3

Practice Activity: Can you start finding these numbers in the world around you - in your room, in your home, while riding in a car or other places?

Let's practice

Which pictures have 1, 2 or 3 birds?

Have you ever seen birds like these? One of our next books will teach you how to know what kind of birds they are. Won't that be a good skill?

Telling symbols apart

Let's add an important skill so it is easier to tell symbols apart. Here is one way: does the symbol have curved or straight lines?

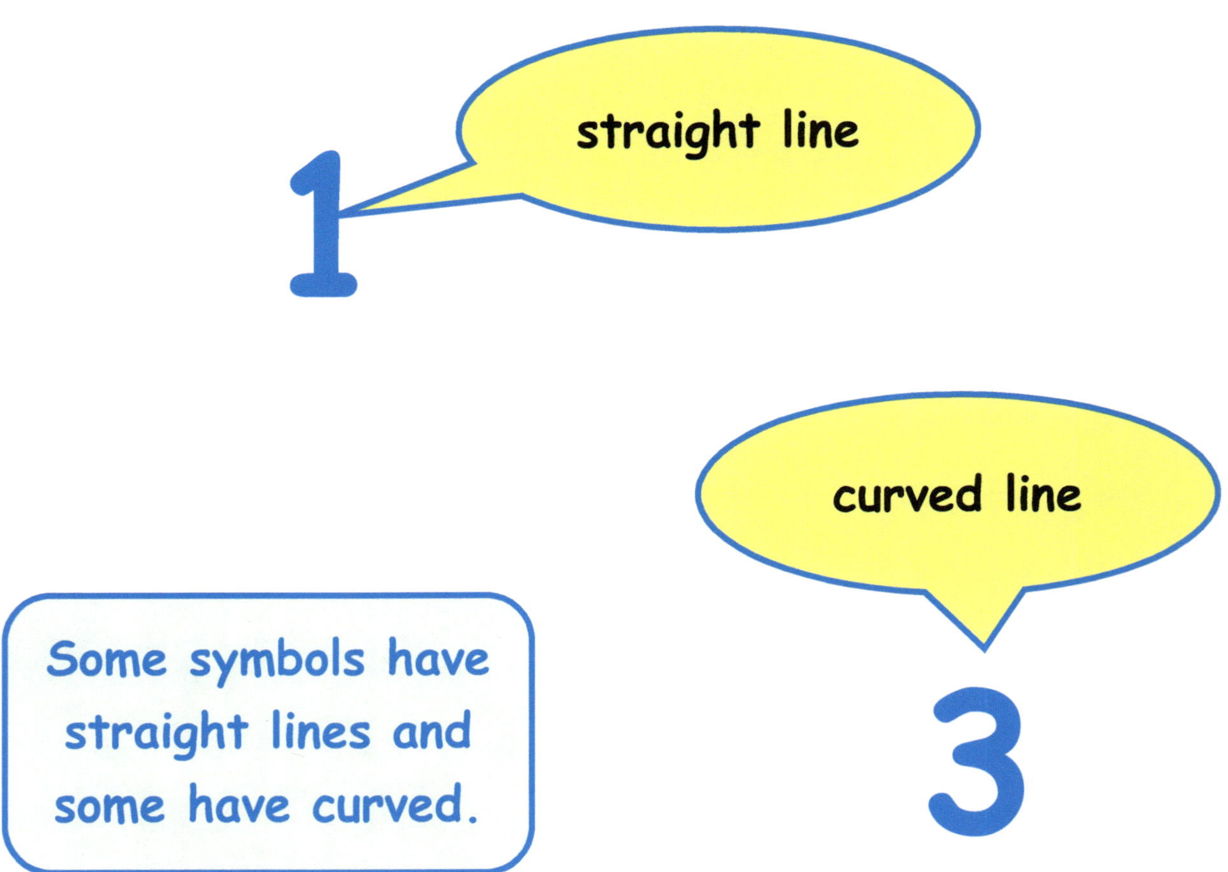

Some symbols have straight lines and some have curved.

How are these symbols alike?
How are they different?
Do you remember what they mean?

4 sea nettles – don't they look like jelly fish? Watch out – they sting!

How many have blue in them?

What else has 5?
Hint: you can wiggle them.

PUMPKINS
6 orange pumpkins
Aren't they big?

$ 6

Now, let's practice!

How many baby chicks are there?
Which symbol is it – 4, 5 or 6?

Hint: If you are stuck, you can look back in the book and see if there is a page that can help you.

Telling symbols apart

Let's learn more ways to tell symbols apart
- by using top, bottom, left and right

top

left **6** right

bottom

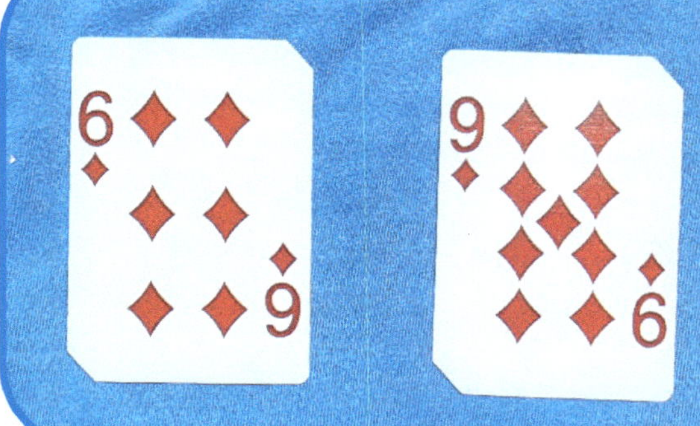

The numbers are in the top left. Can you point to where that is? Can you tell how they are different?

Let's start learning numbers that look like other numbers.

7

7 crazy looking pumpkins!

Doesn't a 7 look like a 1? How is it different? Hint: compare the lines

$ 7

Which is your favorite?

Doesn't an 8 look like a 3? How do you tell which it is?

Hint:
Compare
the curves

9

How do you tell a 9 from a 6?

$ 9

Practice

How many paddles are there - 7,8 or 9?

How many stripes are there - 7,8,or 9?

Time to review

Here are all the number symbols you need to know in order to count

0 and

1	2	3
4	5	6
7	8	9

What about bigger numbers?

What if there are more than 9 items?

What do we do?
Got any ideas?

10

We put the symbols together
to make bigger numbers.

Can you find 10 roseate spoonbills?

11 and higher

Here are some after 10. Can you see a pattern?

A pattern is something that repeats.

What number do you think will be next?

What was your guess?

Did you guess 15?

If you said 15, that's wonderful! If not, you will do better next time. Growing your skills takes lots of practice.

Using our new knowledge

Now that you know your number symbols, let's see how they help us and practice using them.

How can we use numbers?

We can use numbers to play games. These are playing cards. Does the number symbol match the amount of items on the cards?

We can use numbers to compare items? How many parrots - 3 or 4? How many frogs are there - 5,6 or 7?

Let's practice!

Here are some game pieces. There needs to be one for every player. How many people can play?

Which number is that - 1,2,3,4 or 5?

We use numbers for money? How many $ are there - 1,2,3,4 or 5?

Another purpose for numbers

Sometimes, numbers are used to tell things apart like these cars or players on a team.

Have you seen something like this in the world? What do these numbers do?

What if we wanted to know which panda is which? What might we do?

Is there an easy way to know which is which? Or, what if we wanted to talk about just one of the pandas? How can you tell the pandas apart?

Number the pandas

What if we were looking for panda number 2?
Can you find that panda?

What a mess...

How can we make it easier to know how many?

Organizing the pile

We sort them by color. Isn't this so much easier to count?

Remember: Although you might not be able to count all of these cereal loops right now, keep practicing. It will get easier.

What else can be sorted?

You can look back to an earlier page if that will help you count.

Now, let's apply our new knowledge to money!

Can you tell how many dollars you have?

Using actual dollars to practice your numbers can make learning math and money easier when you get older.

A new money skill ...

Can you find the number symbol on this money? What do you think it means?

Did you guess that it is the same as

If you got it right - good for you! If not, you will do better next try. You are learning so much and doing so well!

For Parents

Helpful hints

Primary goals of text:

Introduce how to learn a symbol language and build symbol recognition skills
Build Preschool/Kindergarten readiness skills of counting and sorting
Introduce concepts that help build problem solving and intangible math skills
Introduce skills that make learning more complicated math easier
Build life skills by applying concepts to money and other real world scenarios

Research says:

A research study of 16,000 children found that early math skills were one of the most consistent predictors of later success in school.

Additional research also shows that youngsters who read printed material with an adult have more success with comprehension.

Importance of practice for reinforcement and growth

Becoming successful at math works very similarly to becoming successful at a sport or hobby. It takes lots of practice to grow strong skills. Most students don't get enough of the right practice and it shows up later.

Every child is different

Every youngster has different strengths and every brain makes different connections at different paces. Some youngsters will need more time and repetitions to understand several of the concepts. Take your time. Enjoy teaching your child these important topics. Have lots of laughs – humans can laugh, computers can't.

Encouragement works wonders

Judging can be very counterproductive when it comes to empowering youngsters to learn. Have fun! Stay positive and encourage practice to improve skills. You are the difference in how your child will approach learning. Even if schools can't individualize learning, you can.

How to assess

The great part is that assessing a youngster's early learning can be done simply by observing behaviors and asking good questions. Here are the key skills and what to look for. Please remember every youngster grows different skills at different times and different paces.

Preschool (and life) readiness skills:

(for parents and mentors giving their youngsters skills for today's world from **Easy Math Enrichment** for busy parents)

Counting

Starting with 1 to 5. Help youngsters make the connection to counting and an amount of items. Work up to 10, then 20 and so on. Can they start to see a pattern?

Can they count to 5, then 10, etc. Can they count items?
If they start to notice the pattern, that is great but not required yet.

Comparing

Start with colors and shapes. Then, compare symbols. What features are the same and what are different? We will add measuring concepts like more, less, bigger, smaller, longer, shorter, and other similar comparisons in a different book. Can they begin to apply to the world around them including money?

Can they identify different shapes, colors, numbers?
Can they begin to tell which is more, less, etc

Sorting

Organizing items by shape, color or other features which includes understanding shapes, colors and similar.

Can they organize things by color or shape?
Creating rows and columns would be an advanced version of sorting

Symbol Recognition

Just as in learning to read, math is a symbol language. Connecting symbols to their written words and purpose is embedded in almost all later learning.

Can they match the number symbol with an amount?
(Adding and subtracting between 1-5 will be in another book)

More activities and good questions

Practice makes the difference:

Count everything – fingers, toes, cereal bits, socks and more

Play find the number symbols – in other books, in your home, at the store, on a menu

Play card games with plain cards to help youngsters connect symbols with amounts rather than pictures: Go fish, war, rummy, crazy 8's and others

Play other games that require counting and symbol recognition: both board games and video can be beneficial – numbers are everywhere, math is everywhere

Use practice money to count. See if they can find the $ symbol in the world.

Create visual reminders and examples. Feel free to copy individual pages and post them in familiar places: their bedroom, the fridge and other places.

Ask good questions to foster critical thinking and make connections to the world around them

As you are reading, ask if they know where they have seen some of the items in the pictures, or if the pictures give them ideas of what they can count during everyday activities

During drive time, dinner time or elsewhere, ask if they can find number symbols in the world? Then, ask what's their purpose?

While sharing this book, ask more questions

Reinforce the curved vs straight concept by having them find more curved of straight lines in other numbers. Remind them they can refer back to earlier pages. Feel free to print or copy pages to post in their room or on the fridge.

Reinforce top, bottom, left and right by asking them about locations in pictures or on pages. Can they think about other things that have a top and bottom or right and left.

Ask about colors, shapes, what else they can see in order to build strong observation skills.

As they learn their letters, see if they can find those letters among the pages. The good part is that youngsters don't need to learn all the letters of the alphabet in order to identify the words for 1-5. They only need to know 10 letters, much easier to learn than 26.

About Auntie

Kathy Stoughton M. Ed., aka Auntie G 123, has created an easier way to learn math. She spent years working one on one with students to understand the key issues impacting their success and design strategies to help them. (Many of her students now have careers they never thought possible before working with her.) What she found was that for her students with math troubles, their problems started young. They didn't learn the right skills early and it affected all their learning afterwards.

Now, Kathy creates materials to help families have fun creating more success in math and all the parts of life that math impacts. Thank you to everyone who as spread the message about her books.

More about Kathy's credentials and experience can be found on her website: AuntieG123.com.

Other books from Kathy

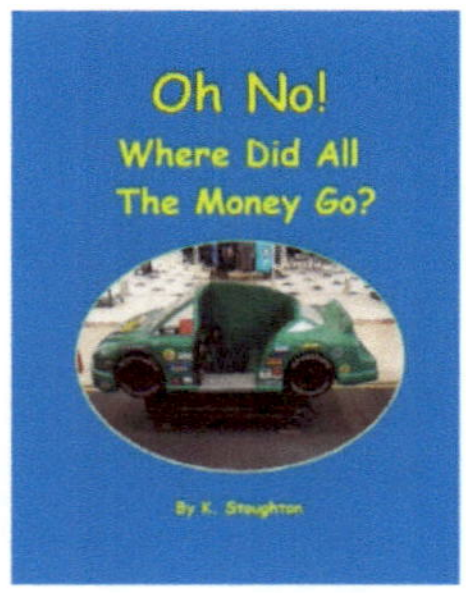

Oh No! Where Did All The Money Go?

Introducing youngsters to more math and money skills including the decimal point to tell dollars from cents, whole from parts and comparing amounts.

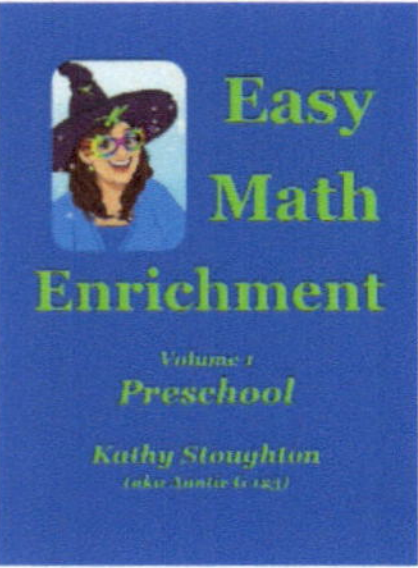

Easy Math Enrichment for Busy Parents (e-book)

Help for parents preventing the troubles they see older students and adults having. The text includes easy to do activities and how to assess progress.

www.ingramcontent.com/pod-product-compliance
Lightning Source LLC
LaVergne TN
LVHW070157110826
845147LV00002B/428
* 9 7 9 8 9 8 8 5 5 0 3 1 0 *